Level= 8

Quiz # = 75465

SCIENCE ON THE EDGE

ARTIFICIAL INTELLIGENCE

• • • •

WRITTEN BY
PHILIP MARGULIES

BLACKBIRCH®
PRESS

THOMSON
GALE

San Diego • Detroit • New York • San Francisco • Cleveland • New Haven, Conn. • Waterville, Maine • London • Munich

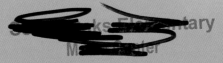

Photo credits: cover © Nick Koudis/Getty Images; pg. 4 © Eriko Sugita/Reuters/Landov; pg. 17 © Issei
Kato/Reuters/Landov; pg. 33 © Susan Goldman/Bloomberg News/Landov; pg. 5 © A. Pasieka/Photo
Researchers; pg. 22 © Sercomi/Photo Researchers; pgs. 32, 41 © Sam Ogden/Photo Researchers; pg. 38 © G.
Tompkinson/Photo Researchers; pg. 6 (top) © Giraudon/Art Resource, NY; pg. 6 (bottom) © Erich
Lessing/Art Resource, NY; pg. 7 © Image Select/Art Resource, NY; pgs. 8, 12 © Hulton|Archive/Getty
Images; pgs. 8 (bottom), 10-11, 26 © Bettmann/CORBIS; pg. 23 © David Woods/CORBIS; pg 28 © CORBIS;
pg. 36 © Reuters NewMedia Inc./CORBIS; pg. 37 © Dale O'Dell/CORBIS; pg. 21 © Bernie
Nunez/Allsport/Getty Images; pg. 23 ©David McNew/Newsmakers/Getty Images;pg. 25 © Douglas
McFadd/Getty Images; pg. 27 ©Bill PierceTime Life Pictures/Getty Images; pgs. 29, 40 © Volker
Steger/Science Photo Library; pg. 39 © Science Photo Library; pg. 35 © Orion/The Kobal Collection; pg. 14
© Universal/The Kobal Collection; pg. 15 © MGM/The Kobal Collection; pgs. 13, 16, 19, 22, 34, 44 ©
PhotoDisc; pg. 18 © NASA; pg. 31 © Anti-Gravity Workshop; pg. 43 © Web Lab (www.mrmind.com)

LIBRARY OF CONGRESS CATALOGING-IN-PUBLICATION DATA

Margulies, Phillip, 1952-
   Artificial intelligence / by Phillip Margulies.
     p. cm. -- (Science on the edge)
   Includes index.
   Summary: Discusses the definition of artificial intelligence, the development of
"thinking" machines, and what computers may be able to do in the future.
   ISBN 1-56711-783-X (hbk. : alk. paper)
   1. Artificial intelligence--Juvenile literature.   [1. Artificial intelligence.]   I. Title. II.
Series: Science on the edge series.

Q335.4.M37 2003
006.3--dc21
                                                                    2002013160

Printed in China
10 9 8 7 6 5 4 3 2

INTRODUCTION

HUMANITY'S COMPETITION

Human beings are not the fastest, strongest, or toughest animals on Earth. They are best in just one thing—thinking. People specialize in thinking just as horses specialize in speed, elephants in strength, and lions in hunting. When it comes to intelligence—the ability to reason, invent, and use what has been learned in the past to plan for the future—humans have no equal.

Some people wonder whether there might be intelligent beings elsewhere in the universe. Since the 1960s, astronomers have aimed radio telescopes at the skies to look for signs of intelligent life on planets many light years away. Whether scientists will someday find intelligent beings, no one knows.

Long before intelligence is found in outer space, chances are it will be created right here on Earth, in the human invention known as the computer. For more than fifty years, computer scientists have been at work on programs that may one day give machines humanlike intelligence. Within a few decades, these scientists will probably achieve their goal. They will make computers that are equal or superior to humans

In the future, intelligent robots like this Japanese prototype may become a part of everyday life.

Artificial intelligence will likely equal or surpass human intelligence in the future.

in every measurable aspect of intelligence. People will talk to these computers directly or on the Internet. Intelligent robots that walk and talk as well as human beings do will be as commonplace as cars are today. When this happens, the human monopoly on intelligence will end, and the era of artificial intelligence will begin.

CHAPTER 1

HISTORY OF ARTIFICIAL INTELLIGENCE

Artificial intelligence is the ability of machines—specifically computers—to think at least as well as human beings do. To be

Blaise Pascal (above) invented the first calculator.

considered intelligent, a computer would have to be able to reason, to learn from experience, to set goals for itself, and to change to meet new conditions.

The ancestors of today's computers first appeared centuries ago. In 1642, French philosopher and mathematician Blaise Pascal invented a calculating machine. Pascal was the son of a tax collector, and he created his machine to help his father count taxes. Pascal's machine was able to add and subtract. Inspired by Pascal's example, in 1643, philosopher and mathematician Gottfried Leibniz invented a more sophisticated machine that used pulleys and gears to add, subtract, multiply, and find square roots. Because Pascal's

Pascal's simple calculating machine began the evolution to artificial intelligence.

and Leibniz's machines only did arithmetic, today they would be considered calculators, not computers. Still, they led the way toward the invention of the computer and, ultimately, artificial intelligence.

THINKING MACHINES IN HISTORY

Leibniz's success with his calculating machine made him wonder why other kinds of thinking—besides arithmetic—might not be mechanized, too. He thought the first step would be to invent a perfect language for a machine to use, in which each word would have only one meaning. Leibniz saw that all human languages had words with many different meanings. He believed that multiple meanings and faulty logic led to errors. A machine that used its own perfect language logically would not only think, but would never make mistakes, he reasoned. Although seventeenth-century technology was not advanced enough for Leibniz to build his machine, his idea for a perfect language anticipated the computer languages that programmers use today.

Gottfried Leibniz (above) envisioned a computerlike machine language.

The next great milestone in the invention of the computer came in the 1840s, when English mathematician Charles Babbage almost made a computer in the modern sense. By the nineteenth century, industry and government needed huge numbers of mathematical calculations to be done all the time. Thousands of people worked hard to do these calculations. The work required skills but was repetitive and boring.

Charles Babbage designed two advanced computing machines.

To take over this job and free people for other work, Babbage designed two computerlike machines. The first, called the "Difference Engine," did advanced mathematical calculations. Although Babbage put together a working model of this machine, he could not raise enough money to finish the project. Babbage also came up with an idea for a more ambitious machine, the "Analytical Engine." This machine would have had almost all of the main elements found in a modern computer, including different kinds of memory

Babbage's Difference Engine performed advanced mathematical calculations.

and the equivalent of a program. The Analytical Engine probably could not have been built with the relatively crude mechanical technology of the mid-nineteenth century, however.

A hundred years later, the need for calculations was greater than ever. The world was ready for computers. In fact, they were invented by people who worked separately—and often secretly—in Germany, England, and the United States during the 1940s. To this day, historians argue about which inventor actually began the computer age, but most agree that by 1951, the computer as it is known today had been invented. Computers have since become faster and more powerful, but they are all still the same basic machine.

All computers have a central processing unit, or CPU. This is the brain of the computer, the part that does calculations. All computers use programs—sets of specific step-by-step instructions that tell the computer what to do. Computers also have a long-term memory, called the hard drive, where programs and data are stored. In addition, computers have random-access memory, or RAM, which they use to store temporary data.

Most of a computer's major parts have equivalents in the human brain. This is not surprising, since computers were designed by humans to do the jobs that human brains do. Like computers, human brains have a kind of central processing unit, where high-order thinking is done. Human brains also have long-term memory (the equivalent of a computer's hard drive) and a short-term, working memory

A computer's hard drive (above) functions is like a person's long-term memory.

(roughly similar to RAM). One big difference between humans and computers, though, is that human beings do not run on programs. Rather than work from exact instructions, human minds grow—they learn from experience and set their own goals.

COMPUTER RESEARCH BEGINS

From the time they first appeared, it was obvious that computers were special machines. They did not bale hay, wash dishes, or do any of the physical tasks other machines did. Instead, their job was to think. It did not take long for people to wonder whether computers would one day be able to think in the same way that human beings do. In 1950, the philosophy magazine *Mind* invited computer pioneer Alan Turing to answer the question "Can machines think?" Turing predicted that computer intelligence would equal human intelligence by the year 2000.

In the 1950s, the United States led the way in computer research.

From the 1950s on, the most advanced computer research was done in the United States. The U.S. government saw that computers could have important uses in warfare; for example, they could aim missiles and respond quickly to surprise attacks. The government gave money to universities to encourage computer research. Some researchers worked to develop programs that would help computers meet immediate military needs, such as an early warning system that would alert the armed forces to a sneak attack. Other scientists tried to find a way to make computers imitate human thought. They wanted to see if computers could learn to

speak and understand English, to tell a story, or to see and hear as humans do. Perhaps computers could become creative enough to come up with unexpected solutions to problems—solutions their designers had never thought of.

In 1956, the pioneers of artificial intelligence (AI) research held a conference at Dartmouth College to bring together AI researchers and to announce that a new science of "artificial intelligence" had officially been born. Teams of researchers then focused their efforts on getting machines to show intelligence in one task at a time. Some researchers tried to teach computers to play games. Others worked to teach computers to solve advanced logical problems. Others helped computers use human languages.

AI researchers realized that computers could not imitate human thought.

As they set out to teach machines special skills, AI researchers were well aware of the obstacles in their path. Although computers were capable of many tasks, they also had certain limitations that did not allow them to think in exactly the same way humans do. The goal of AI research was to find a way to use the power of computers to overcome those limitations.

THE LIMITATIONS OF COMPUTERS

Computers can do amazing things. Even the first computers could calculate with a speed and accuracy that would be impossible for any human being. Computers are able to aim rockets and guide missiles, since in seconds, they can do math that humans would take years to do. Computers also have an inhuman ability to remember facts and find them quickly. It is this ability that allows a computer to bring up thirty-two thousand records on the World Wide Web in just seconds when an Internet user asks it to search for the words "George Washington." What one computer knows can also be taught to another computer in minutes. All someone has to do is load information from the first computer onto a disk and let the second computer read it.

Computers also have limitations, though, and early AI scientists knew they would have to get past these if computers were ever to be intelligent. For example, in the 1950s, computers did not learn from experience, as people do—and most

Today's computers provide Internet users with a wealth of information.

DREAMS OF ARTIFICIAL INTELLIGENCE

The idea of artificial intelligence has fascinated the human race since the invention of the first machine. In Greek mythology, the god Hephaestus built a bronze warrior that ran on steam power. In modern times, the most famous story of artificial intelligence is Mary Shelley's novel *Frankenstein*.

Dr. Frankenstein's artificial creation

In the 1819 novel, a scientist creates an artificial man who rebels against its creator and kills everyone the scientist loves. Since then, the word *Frankenstein* has come to refer to a creation that turns on its creator. Often, machine intelligence has represented the dark side of scientific progress, especially the feeling that it has gone beyond human control. In Samuel Butler's 1872 novel *Erewhon*, a perfect society of the future forbids the invention of new machines, specifically in order to prevent the development of artificial intelligence.

In 1920, Czech writer Karel Capek introduced the word *robot* in a popular stage play, *R.U.R.* Although the word has come to mean a mechanical person, the robots of Capek's play were flesh-and-blood creatures. Like Frankenstein's monster, the robots rise against their masters, and in this case, they win, and replace the human race.

During the 1940s, science-fiction novelist Isaac Asimov decided that it would be refreshing to imagine a friendly relationship between human beings and robots. Asimov's robots were built to put human welfare above their own. Despite Asimov's innovation, most writers and filmmakers continued to be fascinated by the idea of rebellious robots.

From the 1960s on, fictional treatments of artificial intelligence began to reflect an awareness of the kind of AI research that was actually in the works. In Stanley Kubrick's movie *2001: A Space Odyssey*, the spaceship computer H.A.L. 9000 decides that the success of its mission requires it to kill the human crew members. In John Carpenter's *The Terminator*, a computerized expert system that controls nuclear weapons develops self-awareness and decides to destroy the human race. In Steven Spielberg's *Artificial Intelligence*, a race of compassionate robots replaces humans—and the film seems to take the view that this is for the best.

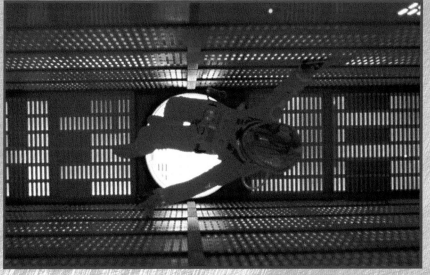

The H.A.L. 9000 computer of *2001: A Space Odyssey* turned against astronauts.

computers still do not. A boy who plays a computer game for the fifth time plays better than he did the first time. He will continue to get better until he is too good for the game and it becomes boring to him. The machine, on the other hand, will play no better than it did the first time. The boy learns from experience, but the computer does not. To teach computers to learn from experience has been a major goal of AI research.

Another limitation of computers is that they cannot guess. They do exactly what they are told to do—and the results can be irritating to humans. An Internet search for George Washington will bring up many websites related to the life of the first president, but it will also turn up sites about George Washington University, George Washington Baked Beans, Washington Avenue, and other subjects that have nothing to do with what the user wants. A human being would have guessed that if a person asked for Web

Children's ability to learn through experience allows them to master computer games.

pages on George Washington, and did not add "University" or "Baked Beans," he or she was probably interested only in pages about the first president. There is a chance that the human being might be wrong, since guesses can lead to mistakes. Still, guessing saves time. Because human beings can make quick decisions based on what has worked for them before, many tasks that they can do quite quickly would take even a very fast computer a long time to do.

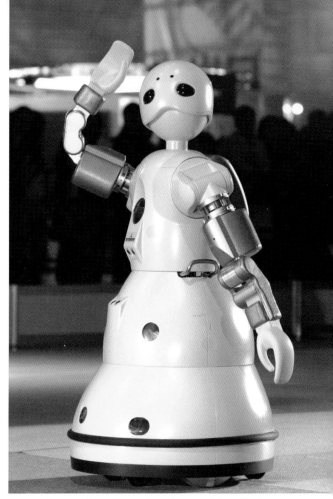

Researchers hope to advance AI so that robots like this one can someday fully communicate in human language.

Another limitation of computers is that they do not understand human language. When humans talk, the meaning of what is said often depends on the situation. For example, the sentence "I hate pepper" has one meaning if another speaker has just said, "I hate horses." It means something else if the other speaker just said, "Would you like some pepper?" It has yet another meaning if the other speaker just said, "Marvin Pepper wants to go to the movies with us." A great deal of complicated and subtle judgment is needed to understand statements in context. AI researchers hope to teach these skills—and many others—to computers.

IS ARTIFICIAL INTELLIGENCE POSSIBLE?

For at least as long as scientists have pursued artificial intelligence, there have been those who have doubted it was even possible. Some point to the fact that there are limits on how much the speed and power of computers can be increased. Computers have doubled in speed and power every two years because manufacturers have been able to make computer parts smaller and smaller. This process cannot go on forever, though. At the current rate of development, in a few years, the insulators on the transistors that form computer switches will be just a few atoms thick. If they get any smaller, they will not work. Therefore, long before they equal the capacity of the human brain, computers will no longer be able to increase in power. In reply to this argument, some computer scientists say that increases in computer power do not depend on any one technology. When the current technology reaches its limits, other, more powerful methods will be found.

AI researchers work to overcome the limitations of present-day computers.

Other doubters say that there are mysterious properties to human intelligence that computers will never be able to match. Mathematician and physicist Roger Penrose suggested that the human brain may rely on subtle effects at the subatomic level (the level of particles smaller than atoms). AI scientists say that if this is true, research will discover these effects and build them into computers.

Ultimately, the AI debate raises age-old questions about the nature of the mind. Philosopher John Searle argued that even if computers could imitate human intelligence, they still will not have minds in the sense that human beings do. To move symbols around is not the same as to know, Searle has said.

To explain what he meant, Searle suggested that people imagine a man whose only language is English receiving instructions in English to move around Chinese symbols. If he uses the symbols as directed, the man might be solving advanced math problems or writing a novel. On the other hand, he could just be moving the symbols around meaninglessly. He would not know the difference. He would not be conscious of what he was doing in Chinese—that consciousness would exist only in the person who gave him the instructions.

Some computer scientists say that Searle's objection may be valid for traditional computers, but not for neural net computers, which process information in a more humanlike way. Others point out, as Alan Turing did long ago, that human beings present each other with the same puzzle that intelligent machines would present. Consciousness is an experience. It cannot be observed from the outside. People have no way to prove that they are conscious.

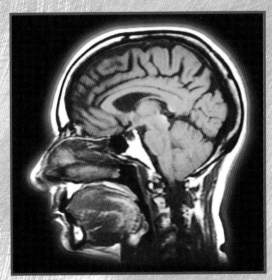

Scientists debate whether computers will ever attain humanlike intelligence.

CHAPTER 2

THE QUEST FOR ARTIFICIAL INTELLIGENCE

Guided by the differences between computer and human intelligence, scientists have broken human intelligence down into separate skills. Each skill—such as the ability to learn from experience, the ability to make guesses, and the ability to understand human (or "natural") language—makes up a different AI research project.

TEACHING COMPUTERS TO LEARN FROM EXPERIENCE

To teach computers to learn from experience, AI researchers devised game-playing computer programs. Games are useful for AI research because progress in game playing is easy to measure. It is simple to test whether a computer has improved at a game. When it gets good enough, it wins, and defeats a human opponent who bested it in earlier matches. Game-playing programs help computers remember the moves and board positions of each game they play, so they can use what they learned the next time they play. Over time, computers play faster, because they do not have to think about the moves they made before all over again. They remember the outcomes of each move they made, and try to use winning moves and to avoid moves that led to losses.

Over the years, AI researchers have also been able to teach computers to use their experiences to make general conclusions. IBM's Deep Blue is a successful game-playing computer. Within the limited domain of chess, Deep Blue has become very powerful. In 1997, it won a match against Gary Kasparov, who is considered the greatest chess player in human history.

Other researchers have used new insights about the way the human brain works to design computers that can learn in a more human way. The electronic circuits inside these computers are arranged to mimic the way nerve cells are set up in the brain. Because nerve cells are also called neurons, these computers are called "neural nets."

From the earliest days of computers, it has been known that the brain is a bit like a computer—neurons use electricity, and nerve cells seem to act as electronic switches. On the other hand, as more has been learned about the brain, some big differences have been found. The circuits in most computers are linked in series. This means they are arranged so that only one action happens at a time. Neurons in the human brain, on the other hand, are linked in parallel. This allows many neurons to fire at once, and, as a result,

An audience witnessed IBM's Deep Blue computer defeat of Gary Kasparov in a 1997 chess match.

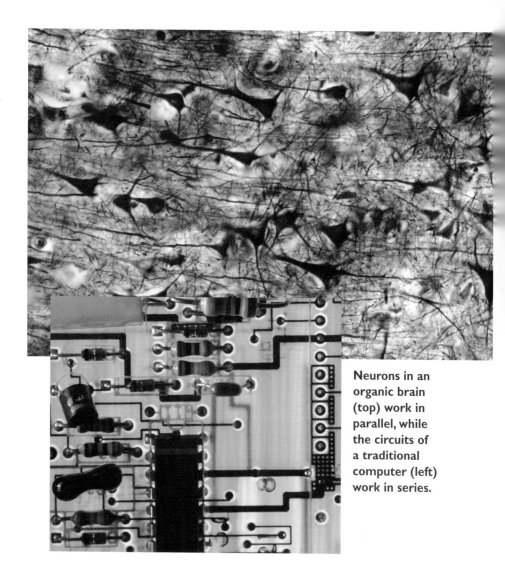

Neurons in an organic brain (top) work in parallel, while the circuits of a traditional computer (left) work in series.

many things may happen at once. The circuits in a neural net computer are also arranged in parallel, like the neurons in the brain. This similarity to the human brain helps neural net computers learn in a way that resembles human learning.

There are many differences between the way computers and human beings learn. Human beings often learn by trial and error. Unlike a computer, a baby does not memorize a complicated set of instructions to learn to walk. Instead, the baby makes guesses and learns through its mistakes. As it tries to walk, the baby moves one muscle, then another. Sometimes the baby fails, and sometimes it succeeds. Every move the baby makes is directed by its brain and is accompanied by some connection between a group of

A baby learns to walk through the process of trial and error.

neurons in the brain. When something works, the brain connection that made the successful move is kept open for later use. When something does not work, the brain shuts down that connection. After a while, the neurons in the brain make the connections that let the baby direct its legs to walk.

Rather than follow specific programs in order to process information, as most computers do, neural net computers learn as babies do, by trial and error. When they make correct choices, they are rewarded—that is, the connections used to make those choices are kept open for future use. When they make wrong choices, the connections are "punished," or shut down.

Robots based on neural net computers learn through experience and do not have programmed instructions.

ROBOTICS: GIVING A MACHINES A BODY

In nature, living things do not sit and wait to receive instructions to tell them what to do. They struggle to survive in a complex, changing world. The organisms that move around most tend to be the most intelligent. Clams and coral do not move, and they have primitive nervous systems. Mammals are highly mobile, and they have complex nervous systems.

Biologists believe that intelligence goes with mobility for a couple of reasons. For one thing, to manage movement requires a complicated nervous system. Beyond that, organisms that move have to be adaptable. They meet new problems as they move through different environments, so they have to develop general problem-solving abilities.

These observations have led some AI researchers to wonder if mobility will help machines achieve intelligence. A machine that travels on its own, rather than sit on a desktop, would face unexpected situations. It would have to exercise problem-solving skills of a more general, flexible kind than those used by a chess-playing machine. For these researchers, robotics—the design of machine bodies—is critical for the development of artificial intelligence.

Meanwhile, engineers who develop robots for practical purposes—from cleaning floors to spying on enemies in battle—find that mobility does indeed require intelligence. Robots that are bolted to the floor on auto assembly lines do not need to be very smart. Robots that wander around an apartment building and automatically vacuum the

An effective robotic vacuum cleaner must have the intelligence to distinguish waste from nonwaste.

floor, however, will need much more complex programming. Even a robot that has just one job to do will have to make decisions about the unexpected—such as whether to vacuum the laces of somebody's shoe.

The computer learning skills developed for neural net computers are widely used today in a class of computer programs called "expert systems." Designers of expert systems collect specialized knowledge from human experts, such as doctors or lawyers, and then program that knowledge into computers. Expert systems use the information to solve problems and make the kinds of decisions human experts would make. Expert systems learn from experience and get better at what they do the longer they are on the job.

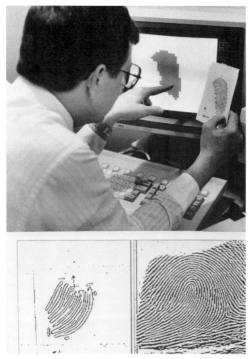

Expert systems help law enforcement with tasks like fingerprint identification.

Fingerprint identification is one job that has been made easier by expert systems. Police need to compare fingerprints found at crime scenes to the hundreds of thousands of fingerprints on file across the country. The job requires an ability to see patterns as well as a high degree of judgment. It also calls for an inhuman resistance to fatigue. It is therefore a perfect task for artificial intelligence.

The detection of credit card fraud is another job that calls for judgment and also requires sifting through mountains of information. Artificial intelligence systems have been developed to do this work, too. These programs scan credit card transactions for anything out of the ordinary, and they alert managers to any suspicious activity.

TEACHING COMPUTERS TO GUESS: HEURISTICS

Although some computers have been taught to learn from experience, most computers have another limitation—they do not know how to make educated guesses. This problem was addressed by an early AI program

Herbert Simon (left) and Allen Newell (right) used heuristics to make computers perform more efficiently.

called the Logic Theorist (LT). LT was developed in the 1950s by AI pioneers Herbert Simon and Allen Newell. Simon and Newell wanted to teach a computer to find—on its own—the proofs for a set of well-known mathematical theorems. (A theorem is a rule of mathematics that can be proven from other, more basic, rules. To prove them is highly advanced, creative work.) This particular group of theorems had already been proven, so there was really no need for a computer to do the job again. Still, any machine that could do the proofs would have to be very intelligent. So, Simon and Newell used this task to find ways to make computers more intelligent.

In theory, a computer could have been programmed to do the proofs blindly—to simply try every possibility. In fact, before LT was developed, this was how computers solved problems. To finish the job with that method, however, would have required an extraordinary amount of time—longer than the life of the universe. Simon and Newell's strategy was to give the computer time-saving ways to guess that would allow it to narrow down its choices. This is how human beings deal with the countless decisions they make each day, which would otherwise overwhelm them. Newell and

Simon called the rules for guessing "heuristics." Heuristics limit choices to what usually works, rather than try everything.

Heuristics have made the computer programs used in many fields much more effective.

Heuristics let many computer programs today make rapid decisions. Fingerprint identification systems and systems that detect credit card fraud use heuristics. So do the expert systems that help diagnose and treat disease, predict the weather, and plan airplane schedules. Heuristics have helped scientists begin to overcome one of the most severe limitations of computers. Even so, there remains another problem to get past.

TEACHING COMPUTERS TO UNDERSTAND NATURAL LANGUAGE

To teach computers human language has been one of the toughest challenges faced by AI researchers. To expand the computer's ability to communicate with human beings, scientists have developed "natural language" programs. (Natural languages are simply human languages, such as English or Chinese. Natural language programs give computers the ability to communicate with people—to write and talk—in regular human language. They also include programs that have computers translate speech or documents.)

Many natural language programs take the form of computer programs that pretend to be human beings. Called "chatterbots," these programs ask and answer questions and try to make people think they are dealing with another human being. One of the most famous chatterbots is a program called Eliza, which pretends to a psychotherapist. If someone types, "It's windy today," Eliza might respond, "How do you feel about that?"—a question psychotherapists often ask. Chatterbots rely heavily on tricks to fool the people who interact with them. They are far less intelligent than they seem to be, but they have helped AI researchers learn a lot about the problem of teaching computers to understand human language.

Other natural language programs read and turn printed words into sound to help blind people, and some programs translate documents from one language to another. Others can even translate spoken words. Users speak to the computer, which then creates a printed version of what was said in one or more foreign languages. Natural language software has begun to be used by search engines, the powerful programs that search the Internet in response to key words or questions. Natural language search engines are not very good yet, but they are likely to get better as they get more experience and power.

This German portable computer uses a natural language program to recognize and translate speech.

VIRTUAL EVOLUTION

Biologists believe that all living things today descend from more primitive creatures. Unlike computer programs, living creatures evolve in the manner described in Charles Darwin's *Origin of Species*. As evolution is understood today, plants and animals do not reproduce themselves exactly—there is always some variation in their offspring. The variants that are best able to adapt to their environments survive long enough to produce another generation, and the less well-adapted variants die off. Over time, this process gives rise to new kinds of plants and animals—new species. Scientists believe they can evolve from simple to more complex.

Some people wonder whether computer programs might evolve, too. In the late 1980s, an evolutionary biologist named Thomas Ray designed a program to test this possibility. For a decade, Ray had observed the struggle for survival among species in the Central American rain forest. His program was called *Tierra* (Spanish for "Earth"). It consisted of a "virtual computer" that could run on many different computers and would act as an environment for computer programs called digital organisms. Like real living things, Ray's organisms could reproduce. They would reproduce much faster than biological organisms do, however, so it would be easier to see evolution at work.

Ray expected that he would have to tinker for years to get his project off the ground, and he was surprised by the results. He explained, "On the night of January 3, 1990, the first time that my self-replicating program ran . . . all hell broke loose. The power of evolution had been unleashed inside the machine, but accelerated to [extremely high] speeds. My research program was suddenly converted from one of design, to one of observation."

Ray has since expanded his experiment to a much larger "biodiversity reserve for digital organisms," which will run on the Internet. He and his sponsors hope to evolve digital organisms that can be harvested for commercial use, much as biologists find useful drugs among the species in the rain forest. With the Internet already plagued by man-made computer viruses, Ray and his colleagues want to make sure that harmful digital organisms do not escape from the preserve. They call this unpleasant possibility the "Jurassic Park scenario." To prevent it, Ray has designed his virtual organisms to be able to work only inside his virtual computers.

These *Tierra* images show a computer's programs in active evolution. The skull (top) represents death removing old and defective programs.

ALREADY HERE OR SOON TO COME?

For some researchers, the fact that solutions have been found for many of the limitations of computers means that artificial intelligence is already here and in use. According to some AI researchers, people just refuse to admit it. Whenever a computer does something that only human beings could do before, people often claim that the task did not require intelligence to perform, after all.

Other AI researchers look forward to the day when various AI programs will come together to become something that can do more than any one of them can separately. They look forward to the achievement of a machine intelligence that is in every way equal to or better than human intelligence. Many researchers believe this will happen very soon.

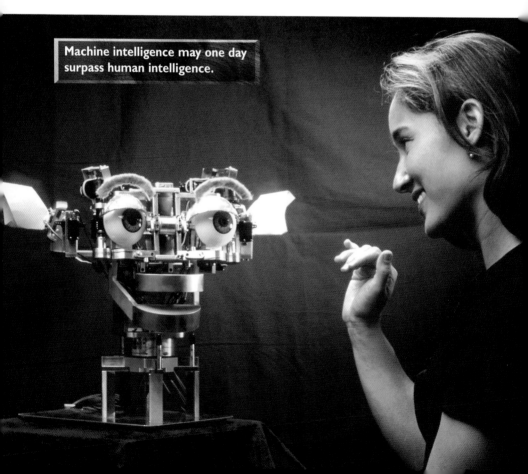

Machine intelligence may one day surpass human intelligence.

CHAPTER 3

THE FUTURE OF ARTIFICIAL INTELLIGENCE

Computer scientists disagree about how close computers are to achieving intelligence. Some think it will take just ten more years. Others believe it is more likely to take another fifty.

On one hand, Toshitada Doi, president of Sony Digital Creatures Laboratory, speaks as if a world of intelligent robots were just around the corner. He has said, "Ten years from now, I believe most households will keep two or three personal robots and their performance will increase 100 times. . . . My expectation is that these robots will be able to talk naturally with humans, say about the latest gossip." Ray Kurzweil, the inventor of AI reading and speech recognition programs, predicts that artificial intelligence will be achieved by 2020. He adds, "We will begin to have relationships with automated personalities."

On the other hand, Rodney Brooks at the Massachusetts Institute of Technology (MIT) thinks that artificial intelligence will take longer to accomplish. He compares AI researchers' current understanding of the nature of intelligence to

If AI becomes fully realized, robots like this one will no longer need human input to operate.

people's knowledge of the solar system five hundred years ago. At that time, astronomers could trace the paths of the planets, but could not explain why the planets moved the way they did. Today, there are basic things about intelligence that scientists still do not know. Until intelligence is better understood, scientists will not be able to give it to a computer.

A DIFFERENT WORLD

Although scientists are unsure of how long it will take to develop artificial intelligence, they are certain of one thing: If and when artificial intelligence is achieved, human life will change a great deal, because machines will be able to do every job that human beings do now. In fact, computers would do most jobs better than humans do, because very soon after a computer becomes as intelligent as an average human being, a computer

Intelligent machines may soon take over tasks much more advanced than those of this present-day welding robot.

will become much more intelligent than any human being. Since their invention, computers have doubled in speed and power about every two years. There is no reason to think that this rate of progress will stop.

Intelligent computers will not just be intelligent, they will be geniuses, and there will be as many of these superintelligent computers as people want. The same technology that makes a personal computer fairly inexpensive now will make superintelligent computers affordable, too.

Today, computers help diagnose and treat diseases. A computer reviews information that has been entered into it and is able to suggest a diagnosis or medication. A computer with artificial intelligence, however, would do much more than just help the doctor. It would replace the doctor, since it would be smarter and would do a better job.

A police robot, like the title character in the film *Robocop* (pictured), could possess superhuman senses.

Because intelligence is part of everything people do, all human activity would be deeply affected by artificial intelligence. A police robot with better than human intelligence could also have superhuman senses (complete with X-ray vision) and a wireless connection to other police robots and to law enforcement databases all over the world. It would be able to match a fingerprint with a name as soon as it arrived at a crime scene—something no human detective could possibly do. A superintelligent machine could choose stocks better than any stockbroker could. AI would change the practice of law because a superintelligent computer would know how to find precedents and apply the law better than any human attorney. Cars with artificial intelligence would

People may someday rely on artificial intelligence, instead of human advisers, to choose stocks.

communicate with each other and move passengers at very high speeds without ever causing an accident. Artificially intelligent factories would make everything—including themselves—so there would be no need for human workers at all. Artificial intelligence could even revolutionize the entertainment industry: Instead of the games and movies that exist today, artificial intelligence could create a dreamworld especially for each individual.

As with other technological advances in the past, some of the changes that artificial intelligence may bring will be beneficial, whereas others will not be so welcome. For example, AI in law enforcement would help catch criminals, but would also pose a threat to privacy. Artificial intelligence could give governments a way to watch every member of society. Today, when people use credit cards or surf the Internet, their actions are recorded, and governments and corporations can learn about people's private lives through those records. In police states—nations that deny their people basic human rights—tracking this information is a way to prevent dissent. Right now, even the most brutal police state does not have enough law officers to sift through all the information available about private individuals. With artificial intelligence, however, a police state could have unlimited numbers of police at its disposal—artificial police.

In the meantime, although artificial intelligence would no doubt create wealth and leisure for the human race, it might also make human beings feel useless, since machines will be able to do virtually anything that human beings can. It might seem pointless to paint a picture or write a book, when a machine could do it better. Some might even wonder why they should learn about the world, when machines know more than any human being can ever know. Even the scientists who created artificial intelligence would become obsolete. Machines would be able to make all the discoveries and inventions the world might need.

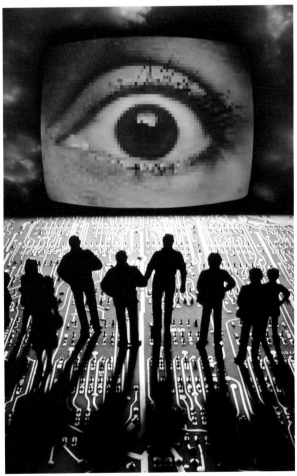

Artificial intelligence could give governments the power to intrude on every citizen's privacy.

"BRUTE FORCE" METHODS: AIRPLANES, NOT BIRDS

Many researchers believe that, although computers might one day surpass humans, they will never think in the same ways.

Many AI applications solve problems in ways that are impossible for the human brain. For example, they take advantage of a computer's quite inhuman capacity to remember facts (Deep Blue, for example, remembered every move that Gary Kasparov ever played in a chess tournament). They also rely on the extreme speed of computer signals—each transistor on a computer chip works about a million times faster than a neuron in the human brain.

AI techniques that make heavy use of these aspects of the computer are said to rely on "brute force" to achieve their results. As computers have increased in power, programmers have had more brute force to work with. To meet their goals, they take advantage of computers' special strengths.

This brings up a point that AI researchers often debate. One day, machines may be able to think as well as or better than human beings, but they will do so in a completely different way. After all, some scientists point out, people do not expect airplanes to fly the same way birds do. One day, computer intelligence may equal human intelligence without resembling it any more than an airplane's flight looks like the wing-flapping flight of a bird.

THE TURING TEST

AI researchers do not agree about how people will know when the age of artificial intelligence has arrived. Many AI scientists disagree about the definition of artificial intelligence. They debate whether it is enough for a computer to be intelligent in just one area or whether a computer must show general intelligence. Some argue that if a computer solves all kinds of problems as well as or better than human beings, that is enough. Others believe an intelligent computer would also have to prove that it was aware of its own existence, as human beings are.

Alan Turing

Mathematician and AI pioneer Alan Turing considered these questions back in 1950, when *Mind* magazine asked him, "Can machines think?" As an answer, Turing proposed a test of computer intelligence. Turing's test is often mentioned when the progress of AI research is discussed. Many people believe that when a computer passes the so-called Turing Test, computers will indeed have become intelligent.

Turing called his test an "imitation game." Three intelligent beings—two human beings and one computer—would take part. For the sake of convenience, the human beings may be called Jane and Fred, and the computer Sam. Jane would be in a room alone and would have a long written conversation with Fred and Sam. At the end of the conversation, Jane would have to decide which of them was the computer. If she guessed incorrectly, then the computer would have passed the test and would be considered intelligent.

Based on the Turing Test, a machine has achieved AI when its responses can pass as those of a human.

Because the participants would be in separate rooms, and the conversation would be written rather than oral, Jane could not guess the truth from Fred or Sam's appearance or voice. Handwriting should not be a clue either, so Turing suggested that the conversation be typed. Today, the Turing Test would be done with keyboards and the Internet. Turing said the conversation should cover many topics, so the computer would not be able to fool Jane with its expert knowledge in just one subject. Jane could also ask Fred and Sam personal questions. She might ask about their love lives, ask them to tell jokes, or ask how they felt about current events.

Turing admitted that the test could not prove whether a machine had the same kind of emotions or thoughts that human beings have. Still, he pointed out that people have the same problem with each other. "The only way to know that a man thinks is to be that particular man," said Turing. "Instead of arguing continually over this point it is usual to have the polite convention that everyone thinks."

A NEW WORLD

From the time human beings started to think about artificial intelligence, they have felt two ways about it. On one hand, it is an exciting idea. On the other hand, many people have felt deeply threatened by it. Both reactions are understandable, since artificial intelligence is bound to bring great changes.

In a way, artificial intelligence is the ultimate invention. Many human inventions have added to the capacities of the body. Knives gave human beings cutting tools that were sharper than animal

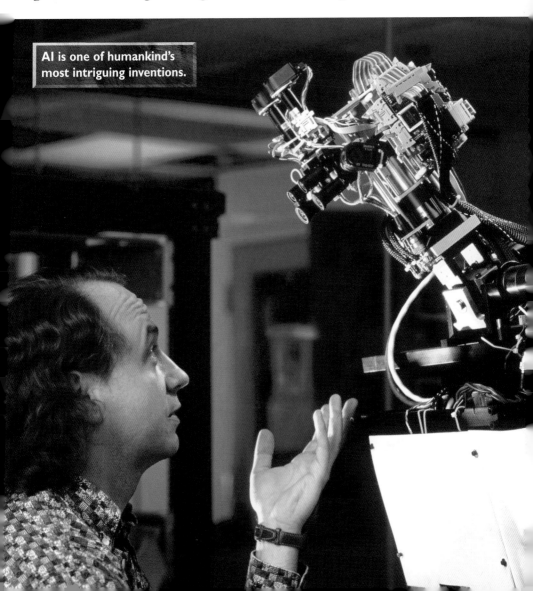

AI is one of humankind's most intriguing inventions.

CHATTERBOTS

A "chatterbot" (short for "chat robot") is an artificial intelligence program designed to carry on a convincing conversation with human beings. Chatterbots are applications of natural language software, usually combined with a certain amount of trickery. People who interact with chatterbots type messages and receive typed responses in return, just as people do in online chat rooms. In fact, people's experiences with chat rooms and instant messages help chatterbots seem more natural.

Each year, new chatterbots compete in a tournament at Boston's Computer Museum. The goal of the tournament is to make human beings think the chatterbots are human. The contest takes the form of a "restricted Turing Test," with rules that make it easier to pass than Turing's original test. The inventors of the winning program receive a cash award called the Loebner Prize.

Chatterbots are usually given a specific job and a personality. One classic chatterbot is Eliza, developed in 1966 by Joseph Weizenbaum at MIT. Eliza is a "computer therapist." She asks and answers questions as if she were in session with a patient. If the user says, "My mother always bakes bread," Eliza will reply, "Tell me more about your family." When Eliza was first introduced, many people formed strong emotional bonds with her. Some psychiatrists even asked Weizenbaum if they could recommend patients to her. Eliza was a fairly simple program that mechanically turned around people's phrases, so Weizenbaum was shocked that people were so easily fooled.

The key to Eliza's success probably lies in the particular role she plays: a therapist. Therapists are not supposed to talk about themselves. They are supposed to get their patients to talk. That is what Eliza does. As a result, she does not have to answer questions that might trip her up. By getting people to talk about themselves, she makes them think about themselves—not about her—so they are less likely to notice her mistakes.

Chatterbots come in many varieties today. For example, Kenneth Colby's PARRY program might be thought of as an answer to Eliza. PARRY is a mental patient—a paranoid schizophrenic. Because people expect his speech to be unusual, PARRY's odd responses are easy to explain. PARRY has fooled psychiatrists in the Loebner tournament.

Web browsers can also access Brian, a computer program that thinks it is an 18-year-old college student. They can also talk with MR. MIND, a chatterbot that addresses itself specifically to the questions raised by the Turing Test. In an application of the rule that the best defense is a good offense, MR. MIND challenges those who interact with it to prove that they are human.

Chatterbots have begun to be used for commercial purposes. Combined with expert systems, they can give out information in a lively, interactive way. For example, Cybion, a French company that publishes information-gathering software for businesses, employs an online "spokesbot" named Cybelle.

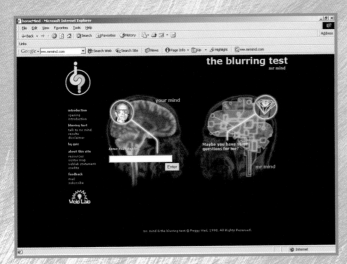

The MR. MIND chatterbot challenges users to prove their own humanity.

teeth. Levers and pulleys added to human strength, and trains and automobiles let human beings move faster than horses. Other inventions expanded the human mind. Writing gave thoughts a permanence that could be passed down to future generations. The invention of mathematics allowed human beings to use symbols to figure out problems too complex for the brain alone. With the arrival of artificial intelligence, the human brain will have expanded its capacity through the creation of a whole other mind—a nonhuman mind, an equal and perhaps a competitor.

Like other inventions, AI will enhance everyday life and challenge human beings to reach even greater heights.

All these developments have changed the way people live, and in so doing, they have changed humanity. Today, humans fly across continents and oceans in airplanes. They communicate via wireless phones with people thousands of miles away, and they move cursors on computer screens to read the words that poets wrote centuries ago. Everyday life today would have been unimaginable to human ancestors, who could not have fathomed the contents of the modern mind. The human race has been changed by its own inventions.

The greatest of these changes will come about in the near future. Artificial intelligence will challenge human beings to think more deeply about what it means to be human. It will challenge us to consider the consequences of sharing a world with beings of our own creation.

GLOSSARY

artificial intelligence computers with human-level intelligence; computer programs that perform tasks once thought to require human flexibility and judgment.

chatterbot a computer program designed to carry on convincing conversations with human beings.

cognitive science the study of the ways human beings learn and process information.

computer chips flat silicon chips microscopically imprinted with the circuitry used in computers.

computer memory the feature of a computer that enables it to recall its past states. Computers have two basic kinds of memory: RAM (random-access memory) is the computer's working memory. Facts stored in RAM disappear when the computer is shut off. ROM (read-only memory) is the computer's long-term memory. Facts stored in ROM are stored until erased by the computer or its user.

computer program a set of instructions given to a computer.

digital computer a computer that processes information in the form of on and off through the use of switches, using binary arithmetic. Almost all computers today are digital.

expert system a computer program that uses specialized knowledge from a human expert and applies that knowledge automatically to make decisions.

hardware in computer science, computers as opposed to their programs.

heuristics rapid decision-making rules in a computer program.

Internet a vast, shifting network of millions of computers linked by telecommunications.

natural language languages used by human beings, as opposed to the artificial languages that are used to program computers.

neural networks machine learning that closely models the physical arrangement and learning processes of neurons in the brain.

programmer a person who creates computer programs.

software computer programs.

transistors electronic components that are able to function both as conductors or insulators of electricity.

Web short for World Wide Web, a portion of the Internet adapted to be used by the general public.

FOR FURTHER INFORMATION

Books

Alison Bing, *Robot Riots: The Good Guide to Bad Bots.* New York: Dorset Press, 2001. A guide to the sport of battle robots, with information about the robots and their builders.

Fred Bortz, *Mind Tools: The Science of Artificial Intelligence.* New York: Franklin Watts, 1992. A detailed look at artificial intelligence. Includes an interview with AI pioneers Herbert Simon and Allen Newell.

Jack Challoner, *Essential Science: Artificial Intelligence.* New York: Dorling Kindersley, 2002. Introduction to the field of artificial intelligence.

Clive Gifford, *How to Build a Robot.* New York: Scholastic Library, 2001. A guide to building a robot, with information about robots in fact and fiction.

David Jefferis, *Artificial Intelligence: Robotics and Machine Evolution.* Ontario, Canada: Crabtree, 1999. An introduction to the field of artificial intelligence and robotics, discussing the history of artificial intelligence and today's use of robots in factories and space exploration.

Websites

American Association for Artificial Intelligence
www.aaai.org

The Computer Museum
www.computerhistory.org

HomeMind
www.mrmind.com/mrmind3
The chatterbot MR. MIND challenges net surfers to prove they are human.

ELIZA – A Friend You Could Never Have Before
www-ai.ijs.si/eliza/eliza.html
A Web-based version of the classic psychiatric chatterbot.

Marshall Brain's How Stuff Works
How Chess Computers Work
http://biz.howstuffworks.com/chess.htm

Marshall Brain's How Stuff Works
How Robots Work
http://biz.howstuffowrks.com/robot.htm

ABOUT THE AUTHOR

Philip Margulies is a writer who lives in New York City.

INDEX